Science Investigations

LIGHT:
AN INVESTIGATION

JOHN GORMAN

PowerKiDS press.

New York

Published in 2008 by The Rosen Publishing Group, Inc.
29 East 21st Street, New York, NY 10010

First Edition

The publishers would like to thank the following for permission to reproduce these photographs:
Alamy: 16 bottom (David Hancock), 28 (Everynight Images); Corbis: 5 (Otto Rogge), 10 (Images.com), 15 (Bo Zaunders), 20 top (Tom Bean), 20 bottom (James Randklev), 26 (Roger Ressmeyer), 27 (Jose Luis Pelaez, Inc.), Cover and 29 (Masahiro Sano); Ecoscene: 22 (Kay Hart); NASA Images: 11; OSF/Photolibrary: 4 (Phototake Inc.), 6 top (Botanica), 6 bottom (Index Stock Imagery), 8 left (David Cayless), 8 right (Michael Leach), 12 (Workbook, Inc.), 14 (Index Stock Imagery), 16 top (Images.Com), 18 (Pacific Stock), 23 (Colin Milkins); Science Photo Library: 13 (Andrew Syred), 25 (Pascal Goetgheluck); Topfoto: 24 (Rosie Scott-Taggart).

The right of John Gorman to be identified as the author and Peter Bull as the artist has been asserted by them in accordance with the Copyright, Designs, and Patents Act 1988.

Editors: Sarah Doughty and Rachel Minay
Series design: Derek Lee
Book design: Malcolm Walker
Illustrator: Peter Bull
Text consultant: Dr. Mike Goldsmith

Library of Congress Cataloging-in-Publication Data

Gorman, John.
  Light : an investigation / John Gorman. — 1st ed.
    p. cm. — (Science investigations)
  Includes bibliographical references and index.
  ISBN 978-1-4042-4286-9 (library binding)
  1. Light—Juvenile literature. I. Title.
  QC360.G676 2008
  535—dc22

                                    2007032609

Manufactured in China

Contents

Why is light so important?

From the time you open your eyes in the morning until the time you close them at night, you can see. Some of the things you see will be close to you, such as this book. Other things, such as the stars in the night sky, will be many millions of miles away. Yet something connects you to both distant stars and the book in your hand, and that something is light. Without light you cannot see. It is light that makes life on Earth possible. If the Sun stopped shining or something blotted out the light of the Sun, the Earth would turn dark and cold. All the green plants would die, because they need light to make their food. So, too, would the plant-eating animals, and in time, every living thing. Fortunately, the Sun is going to continue to shine for many millions of years.

SUN WARNING!
Do not look directly at the Sun: this could damage your eyes or blind you.

The Sun is a star—a massive ball of hot gas. The brighter areas you can see in the photo are explosions on the Sun's surface, known as *solar flares*. The darker patches, which are the cooler areas of the surface, are called *sunspots*. *Prominences* are huge streams of gas that shoot out from the surface into space.

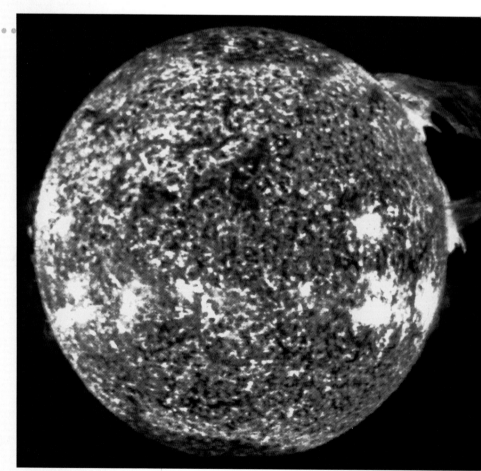

4

INVESTIGATION

How much light is needed to see detail?

MATERIALS

A cardboard tube about 10 in. (25 cm) long, tape, scissors, a pencil, black paper, white paper, and a mouse mat.

INSTRUCTIONS

Ask an adult to cut a slit ¼ in. (.5 cm) wide in the cardboard tube. The slit should be about 4 in. (10 cm) long and 1 in. (2 cm) from one end. Use black paper to make a tube 6 in. (15 cm) long (fastened with tape) that will slide over the cardboard tube, covering the slit. On the white paper, draw three tiny dots very close together. Put the paper on the mat and place the tube over the dots. Press firmly so that light is kept out. Look down the tube with one eye close to the rim. Slowly bring the outer sleeve up until you can see the three separate dots. Measure the length you have made your slit at this point.

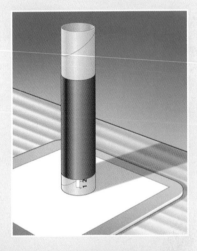

FURTHER INVESTIGATION

Try repeating the experiment with friends and compare the results.

Start with plenty of light and slide the outer sleeve down until you cannot see the dots. Is the length of slit at this point the same as in the investigation?

Try the experiment using dots of color. Do different colors need different amounts of light to be seen?

The stars light up our night sky. They look tiny because they are many millions of miles away from us.

How are shadows made?

A shadow is formed when light rays are stopped from passing through an object.

This stained glass window has colored glass that is translucent and clear glass that is transparent. The strips of lead between the windowpanes are opaque.

Pure water and air are transparent. *Transparent* means that light passes straight through something. Materials that reflect some light, but let some light through, are called *translucent*. Translucent materials are hard to see through. If you are on one side of frosted glass and your friends are on the other, you might be able to make out their shapes, but you would not see their faces clearly. Materials that allow no light to pass through are said to be *opaque*.

Light travels in straight lines. When rays of light strike something opaque, the rays cannot bend around the object in their path. A shadow—the same shape as the object that has blocked the rays of light—is formed behind the object. Your body is opaque and blocks the Sun's rays on a sunny day.

INVESTIGATION

When does the same object have shadows of different sizes?

MATERIALS

A flashlight, a yardstick or a tape measure, a brick, and a wall.

INSTRUCTIONS

Place the flashlight so it shines on the brick. A shadow forms on the wall behind it. Place the yardstick or tape measure so it can be used to measure the distance of the flashlight from the brick.

Set the brick at a distance of 40 in. (1 m) and move the flashlight closer to the brick in 8-in. (20-cm) steps. What happens to the shadow on the wall?

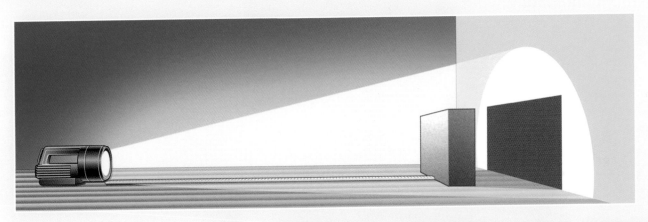

You will notice that the brick has shadows of different sizes depending on how close the light is to the brick. The closer the light source is to the object, the larger the shadow.

FURTHER INVESTIGATION

Record the results in a table and make a bar graph from them. You may prefer to use a computer program to draw the graph. Use the graph to find what the height of the shadow would be at 12 in. (30 cm), 20 in. (50 cm), and 28 in. (70 cm). What happens to the size of the shadow if you move the flashlight and brick farther back?

How do we see?

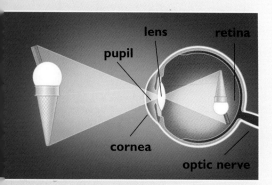

Light shining from an object is detected in the retina and then turned into electrical signals that pass along the optic nerve to the brain.

You see the page you are reading because light rays are being reflected (bounced) off of it and are entering your eyes through your pupils. The light rays are focused by the corneas (the slight bulges at the front of your eyes) and by the lenses in your eyes, which are behind the pupils. The light rays are falling as focused, upside-down images on the retinas, which are the light-sensitive areas at the back of your eyes. The light rays that fall on the retinas are sent as tiny electric messages through the optic nerve to the brain. Your brain changes these messages into a picture of what you are seeing. You look at something with two eyes, but to see two separate pictures would be confusing. Your brain makes sense of the information it receives, and you see two pictures as one image the right way up.

The eyes of this zebra are on the sides of its head. Most animals with eyes in this side position can see almost all the way around including behind. This helps them see predators that attack from behind.

Nocturnal animals have eyes that are adapted to hunting at night. Some, such as the owl and bush baby, have large eyes and wide pupils to collect the light.

INVESTIGATION

Can you see through your hand?

MATERIALS

A cardboard tube about 10 in. (25 cm) long, a piece of letter paper, and tape.

INSTRUCTIONS

Hold the tube in your right hand and bring it up to your right eye. Keep your eyes open.

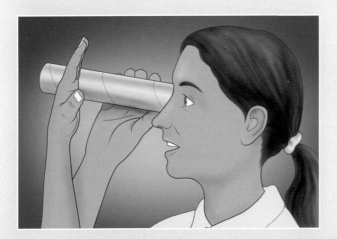

Hold the palm of your left hand, next to the bottom of the tube, as shown in the diagram, and slowly bring it along the tube toward you. You will see a hole in your hand!

Roll the paper into a tube and tape.

Repeat the first two steps but gently squeeze the tube as you bring your hand up—see the different shaped holes you can make.

You have "tricked" your brain by sending it information about two different pictures, instead of slightly different information about the same picture.

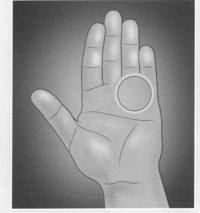

FURTHER INVESTIGATION

Ask a friend to hold a pencil a little more than 20 in. (50 cm) away from you. Try to touch the pencil's point with one eye closed.

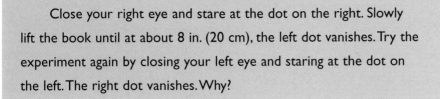

Close your right eye and stare at the dot on the right. Slowly lift the book until at about 8 in. (20 cm), the left dot vanishes. Try the experiment again by closing your left eye and staring at the dot on the left. The right dot vanishes. Why?

Is seeing believing?

Sometimes our eyes can trick us. What you see does not seem quite right or it appears to change as you study it. This impression is called an *optical illusion*. Your brain always tries to make sense of the information it receives from your eyes by matching the picture it makes with previous memories. The brain's struggle to make sense of an unreal picture and its possible meaning creates the illusion. It deceives you into thinking something is completely different from what it is really like.

The film you watch at the movies or on TV is a type of optical illusion. To give the impression of movement, about 24 still pictures per second are projected onto the screen. Your brain sees light for about $1/9$th of a second after it has stopped shining, so fast-moving still pictures merge to create an illusion of movement as a moving picture.

Look closely at the picture. Notice how the corners appear to slant inward and then outward!

INVESTIGATION

Do your friends see identical illusions differently?

MATERIALS

A large piece of white card, a felt-tip pen, a long ruler, a protractor, and a small piece of card.

INSTRUCTIONS

Using the felt-tip pen and ruler, draw the illusion as shown on the large piece of card. Draw the lines at least 12 in. (30 cm) apart.

Ask friends to help with an experiment about optical illusions. Tell them you aren't going to trick them, but that not everyone sees things in the same way.

Ask each person to put the small card over line A and then slide it until the uncovered length of line A appears to be equal to line B in length. Measure and record the uncovered length to the nearest ¹/₂ in. (1 cm).

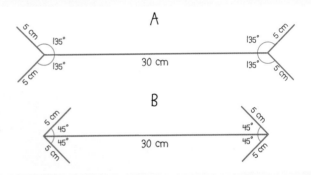

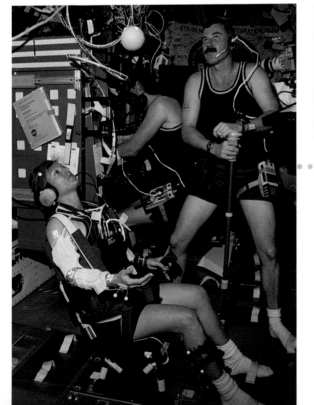

FURTHER INVESTIGATION

Make a graph from the table. If the people you asked were different ages, plot the age of the person against the length of the line perceived. Have younger people seen the illusion differently from older people?

This astronaut is tackling another kind of optical illusion. Astronauts who are new to a gravity-free environment find it difficult to catch balls accurately. Their brains need to learn how this environment affects the way the ball travels. With practice, their brains are able to store this new information and the astronauts can catch the balls accurately.

How does light reflect off of an object?

When light falls on objects, the light rays reflect off of them into our eyes. Some of the things we see shine, because of the way they reflect the light. Look closely at objects that are shiny, because they are not all the same. They may be made from different materials and be of different colors, but they all have one thing in common: they all have smooth surfaces. If a surface is smooth then the light rays are reflected together in the same direction. However, if the surface is rough, then the light rays are reflected in many different directions and the object appears dull.

The smooth surface of the water gives a good reflection of the surrounding scene.

Mirrors have smooth surfaces that reflect the light that falls on them so accurately that we can see ourselves. Mirrors can also let us see around corners if we position them in a certain way.

INVESTIGATION

How do different surfaces reflect light?

MATERIALS

A mirror, a piece of dark card with 1/4-in. (0.5-cm) strips cut out, a pencil, a flashlight, a piece of white paper, a piece of black paper, some colored and clear plastic, a piece of wood, a smooth piece of kitchen foil, and the top of a tin cookie jar.

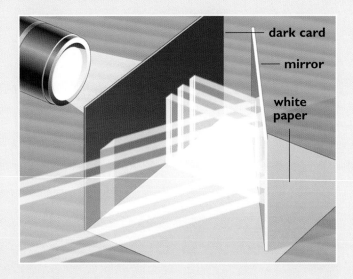

dark card

mirror

white paper

INSTRUCTIONS

Take the materials to a dark area and set them up as in the diagram.

Notice what the mirror does to the rays of light that shine on it. Record the position of the reflected rays on the white paper. Next, replace the mirror with each of the other materials to see how they reflect light.

Record what happens to the rays of light when they shine on the different surfaces. Use the words "reflect" and "absorb" (to take in) to help explain what you have seen.

FURTHER INVESTIGATION

Print "CHOICE" and "BALL" on separate pieces of paper. Hold a mirror along the top edge of each word and study the reflections. Why does one word appear the right way up, but the other appears upside down?

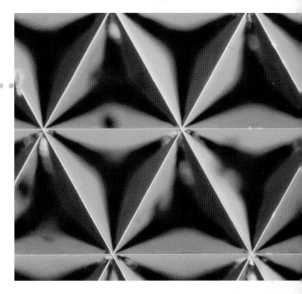

This is a close-up of the reflective material used in safety clothing. Light reflects off of the angled surfaces back to its source, making the wearer much more visible in dark conditions.

What do you see when you look in a mirror?

When you look in a mirror, you see a picture of yourself. Light rays have reflected from your face onto the mirror that has bounced the light back into your eyes. However, this mirror image is not an exact picture of you. Blink at yourself with your right eye and your left one blinks back! Think of a friend standing in place of the mirror, and both of you blinking your right eye at the same time. The friend appears to blink a different eye, until you realize that his or her right is your left.

Not all mirrors are flat—some are curved and these reflect light in a different way. A *concave* mirror, such as a shaving mirror, curves inward. This makes near things look bigger. A *convex* mirror, such as a driver's rear view mirror, curves outward. This makes things look smaller but shows a wide field of view.

If an amusement park mirror makes you look fatter and shorter, would it be concave or convex? If it makes you look longer and thinner, would it be concave or convex?

INVESTIGATION

How tall does a mirror have to be to show your full reflection?

MATERIALS

A long mirror, three pieces of card 16 x 16 in. (40 x 40 cm), 24 x 16 in. (60 x 40 cm), and 32 x 16 in. (80 cm x 40 cm), a felt-tip pen, a yardstick, and a pair of scissors.

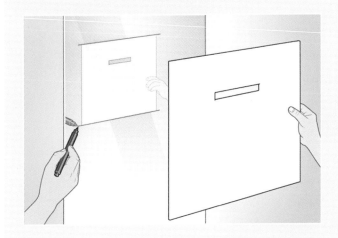

INSTRUCTIONS

Cut out a rectangle large enough to see through, about 4 in. (10 cm) from the top of each card.

Hold a card in front of the mirror and look through the rectangle. Ask a friend to mark where the top and bottom of the card appear on the mirror with the felt-tip pen. Measure and record the distance between the marks. Repeat this process with each size of card.

Study the results: do you notice that each card's mirror image is half of its actual length? What length of mirror could reflect your full image?

The wing mirror has a convex surface that bends light, so the driver can see behind and to the side of the car.

FURTHER INVESTIGATION

What is the smallest length of mirror needed to reflect a man whose height is 6 feet (1.8 meters)? Does a dentist use a convex or concave mirror to look at our teeth? Which piece of cutlery can be used as a convex or concave mirror?

Can you bend light?

When you run from the beach into the sea, the water slows you down because water is denser (thicker) than air. Light also travels at different speeds as it passes through different transparent substances. It travels fastest through space, where there is nothing to slow it down, slower through air, and even slower when it travels through water.

This change of speed when light enters a different substance causes the light rays to alter their direction. Scientists call this change or "bending" of light *refraction*. The refraction of light is the reason why things look so different when they are viewed through water. Notice how the bottom of a swimming pool can look shallow, even at the deep end! The light rays are refracting as they pass out of the water into the air. This tricks the brain, which always thinks that light is traveling in straight lines.

How can a glass bend a spoon? Add water.

A mirage is formed when light is bent, or refracted, by a layer of warm air near the ground.

INVESTIGATION

Can you bend light through water?

MATERIALS

A shoebox, a powerful flashlight, a large piece of white paper, a pencil, a protractor, and a transparent tank (with straight sides) filled with water.

INSTRUCTIONS

Set up a lightbox by attaching a flashlight to the inside of a shoebox. The flashlight should shine through a ¼ in. (0.5 cm) slit at one end. Place the lid on the box.

Place the lightbox and the transparent tank on the large sheet of paper. Position the lightbox so the beam of the flashlight hits the side of the tank at about a 50-degree angle.

Darken the room so you can record the path of the beam. Draw crosses from the lightbox to the tank and from where the beam leaves the tank to the edge of the paper.

Connect the crosses to make two lines.

The two lines show the path taken by the beam: notice how it refracted when it entered and left the water.

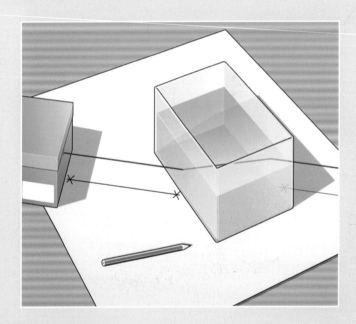

FURTHER INVESTIGATION

Place an opaque bowl on a table and put a coin in it. Ask a friend to look at the coin. Then tell your friend to move slowly backward until he or she has just lost sight of the coin. Add water slowly to the bowl so you do not move the coin. Stop when your friend can see the coin. How does this happen? Think about the way light refracts in water.

What is white light made of?

When you see a rainbow, the Sun is behind you. The Sun's light is reflected and refracted by the raindrops, and the white light is split into the same colors that are produced using a prism.

We talk of white light as if it is a single color, but it is really a combination of different colors. You see these colors when you look at a rainbow or a soap bubble. When sunlight passes through a raindrop, the light splits into the colors that make up the rainbow. Every rainbow has these colors in the same order: red, orange, yellow, green, blue, indigo, and violet. Together they are called the *colors of the spectrum.*

Diamonds are cut to a particular shape so that they refract and reflect the colors of the spectrum. This is what creates the sparkling brilliance of a diamond necklace.

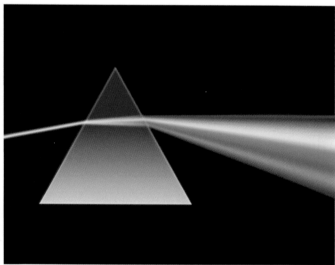

White light can be split into the colors of the spectrum by passing it through a prism, a wedged-shaped piece of glass. The colors separate because each color is being refracted by the prism at a slightly different angle.

INVESTIGATION

Can we make rainbow colors from white light?

MATERIALS

An unwanted CD, a powerful flashlight, and a piece of white paper.

INSTRUCTIONS

Take the materials to a dark area. Place the CD, blank side up, on a flat surface and shine your flashlight at it. Hold the white paper so that the light reflecting off of the CD shines on the paper. Tilt the paper until you see rainbow colors.

At this angle of the flashlight and paper the tiny ridges on the CD have split white light into rainbow colors.

Next hold the CD in one hand. Shine the flashlight onto the blank side of the CD. Tilt it until you see different bands of bright colors on the CD.

These bands are made when light rays reflecting from the ridges on the CD overlap and mix to make stronger or different colors. Such colors are called "interference" colors, since the light rays have overlapped and "interfered" with each other.

Move the CD and flashlight apart. Notice how the bands of color widen. Light always spreads as it travels.

FURTHER INVESTIGATION

Divide a white card disk and color it as shown in the diagram. Attach the disk to an electric motor. When the disk revolves, the colors of the spectrum merge, making an off-white shade. The faster the disk spins, the better the results.

You can use other disks to discover what colors are made when different combinations of two or three colors are mixed.

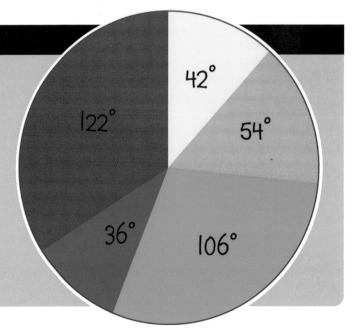

How do objects show their color?

As sunlight passes through the air, violet light is scattered more than blue, which is scattered more than green, and so on. Because there is more blue light than violet light in the first place, and because our eyes are less sensitive to violet than blue, we see the sky as blue.

At sunset, the Sun is low in the sky. The light has taken a longer path through the air as it travels toward us. Most of the blue light has been scattered. More of the red end of the spectrum remains. When there is dust and water vapor in the atmosphere, the red light is reflected in all directions. This causes the brilliant sunsets we see.

EVIDENCE

We see objects when light is reflected off of them into our eyes. The color of an object depends on the color of the light it reflects. A red car reflects only red light. Yet the sunlight that shines on the car is made from all the colors of the rainbow. So what has happened to change sunlight into just red light?

When sunlight shines on an object, substances called *pigments* absorb all the colors of the spectrum apart from their own color. The red car is red because the pigments in the paint have absorbed all the other colors, leaving only red to be reflected. When light falls on a green leaf, the pigments absorb all the rays of the spectrum apart from green, so the leaf reflects green light into the eye. If all the colors of the spectrum are absorbed when light shines on an object, we see black. When an object reflects back all the colors of the spectrum, we see white.

INVESTIGATION

What happens when you mix colored light?

MATERIALS

Three powerful flashlights, pieces of red, blue, and green cellophane, scissors, a large piece of white card. and clear tape.

INSTRUCTIONS

Fit a piece of different colored cellophane to each flashlight and stick it down with the tape. Tape the white card to a wall.

Make the room as dark as possible. Shine pairs of flashlights, one at a time, on the card so that their beams meet.

- What do blue and green light combine to make?
- What do green and red light combine to make?
- What do red and blue light combine to make?

Shine all three colors on the card. What color do they combine to make?

FURTHER INVESTIGATION

In a brightly lit room, use the pieces of cellophane or acetate to investigate the effect that colored light has on the way we see other colors. Look through each piece, one at a time, at different-colored objects. Record the results in a table showing the color of the filter, the colors of the objects, and their color when they are viewed through the filter.

COLOR OF OBJECT	COLOR OF FILTER		
	Red	Green	Blue
White	*Red*		
Red			
Green			
Blue			

Why do plants need light?

Green plants make their own food with the help of the sunlight that shines on them. Each green leaf is like a small factory that uses light energy to get things done. The light energy is used to make food for the plant from water and carbon dioxide, a gas found in the air. If light is blocked off from a plant, no food can be made and the plant dies. All green plants need light, and grow where and when they can find it. Woodland plants flower in early springtime when the trees surrounding them are leafless and the sunlight can shine through to the woodland floor. Light also affects the appearance of a plant. Trees of the same species grown in a crowded wood are taller but have fewer spreading branches than those grown out in the open.

Bluebells grow and flower in the spring, when the sunlight can reach the plants through the leafless branches.

INVESTIGATION

Do plants grow better in white or colored light?

MATERIALS

Compost, five large plastic beakers (pint-size is ideal), rolls of red, blue, yellow, and green cellophane, five plant pots, radish seeds, tape, scissors, and labels.

INSTRUCTIONS

Cover the side and base of four of the beakers with a different color of cellophane. Fill the five flowerpots with compost and place a radish seed at a depth of 1 in. (2.5 cm). Position each of the beakers on the flowerpots as shown in the diagram.

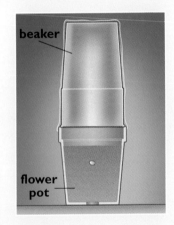

beaker

flower pot

Water the compost and keep it damp. Place the pots in strong light out of each other's shadow.

Predict which seeds will grow into the tallest and the smallest plant.

If you have a digital camera, use it to keep a photographic record every second day. Measure and record the growth using a spreadsheet with columns for dates and heights. Make brief notes about how the plants are growing.

FURTHER INVESTIGATION

Make a display using the photographs. Turn the measurements into graphs using a computer and display them side by side with the notes. Highlight your prediction and the results.

Green plants give off oxygen when sunlight shines on them. Notice the oxygen bubbles on this water plant.

How do we record pictures from light?

If you want to take snapshots of a birthday party, you probably use a film camera and then take the film to be developed. If you have a digital camera, you can take your snapshots and turn them into photographs using a computer and printer.

Digital cameras have lenses that focus light just like a film camera, but in a digital camera, the light strikes sensors not a film. Devices in the camera change the light rays into electrical signals that it can store in its memory. In this photograph the signal has been downloaded onto a computer and the photographs appear on the monitor.

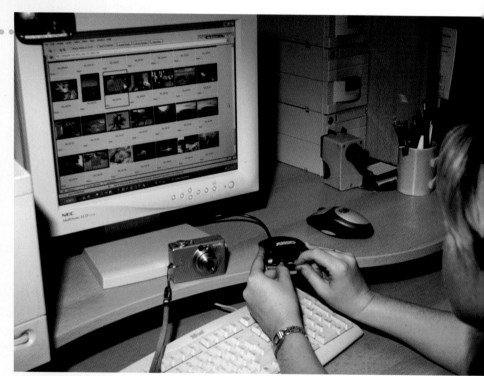

When you use a film camera, light rays pass through the lens and reach the film where they are recorded on the light-sensitive surface. The surface of the film records an image made by the pattern of the light that reaches it. When the film is treated with certain chemicals, the image becomes visible. This process is called *developing*. The image produced by developing is dark where the scene is light, and light where the scene is dark. This type of image is called a *negative*. When light is shined through the negative onto photographic paper, the image is reversed and the original scene can be viewed as a print.

INVESTIGATION

Can you make a pinhole camera?

MATERIALS

A tube-shaped carton with a clear or translucent plastic lid, black paper, a ruler, a thumbtack, strong, wide tape, scissors, and a pencil.

INSTRUCTIONS

Ask an adult to cut the carton along a line 2 in. (5 cm) from the bottom and to make a hole in the base using a thumbtack.

Fit all the pieces together as shown in ii). Use strong, wide tape to hold them together.

Wrap and tape black paper around the sides of the camera to keep light out of it.

Use the camera on a sunny day. Stand in the shade and bring the camera close to one of your eyes. Keep out any stray light and you will see an upside-down colored picture of what the camera is pointing at.

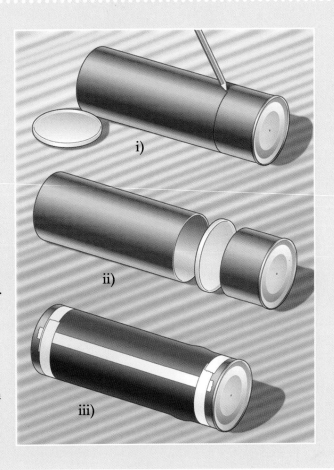

i)

ii)

iii)

FURTHER INVESTIGATION

Light travels in straight lines. How does this help us to explain why the pictures are upside down?

If all camera lenses are circular, how can photographs be rectangular? (Clue: How can you make a rectangle from a circle?)

Red eyes in a photograph are caused by the flash of light reflecting off of the blood vessels in the retina at the back of the eyes.

How do lenses help us to see?

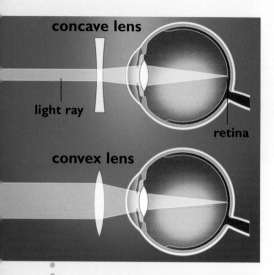

concave lens

light ray

retina

convex lens

In nearsighted people, the light rays focus in front of the retina. In farsighted people, the light rays would be in focus behind the retina. Convex lenses will correct short sight and concave lenses will correct long sight. The lenses bend the light rays so they fall as a focused image on the retina.

The largest refracting telescope in the world has a 40-inch lens. It is at the Yerkes Observatory in Wisconsin.

EVIDENCE

When you look through a magnifying glass or a telescope, you are using lenses to help you see. Lenses are carefully shaped pieces of transparent materials, such as glass or plastic, that make the object you view look larger.

There are two main types of lenses: convex lenses and concave lenses. A *concave lens* curves inward and is thinner in the middle than at the edges. Light beams are spread out by concave lenses and things look smaller when you look through them. A *convex lens*, such as a magnifying glass, curves outward and is thicker in the middle than at the edges. Light rays are bent inward as they pass through a convex lens and they meet, making a focused, clear picture. Both convex and concave lenses are used to help people with eyesight problems. Concave lenses are used in glasses for nearsighted people and convex lenses in glasses for farsighted people.

INVESTIGATION

Can you make a picture with a magnifying glass?

MATERIALS

A piece of white card and a magnifying glass.

INSTRUCTIONS

Find a room that has just one source of light. A window on a sunny day is ideal.

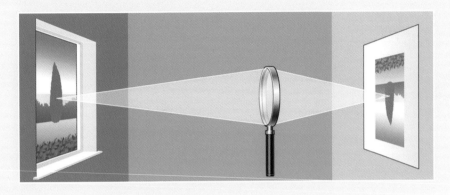

Stand 5 yards (4.5 meters) from the window. Face away from the window and hold the card out in front of you and to the side, so light from the window falls on it.

Position the magnifying lens next to the card and slowly move it away, toward the window, until you can see a clear upside-down picture of the light source on the card.

Move closer to the window and repeat the last step at yard or meter intervals.

FURTHER INVESTIGATION

Is the image of the window larger or smaller than the window itself? What did you notice about the size of the picture when you moved closer to the window?

Why should a magnifying glass be kept out of the Sun?

The girl in this picture is nearsighted and wears glasses for reading. Nearsighted and farsighted people can correct their eyesight by wearing glasses or contact lenses. Contact lenses are thin pieces of plastic placed over the cornea. They float on the thin film of tear in the eyes.

27

How does light help us?

Light has many obvious uses, but it is also a hidden helper. Many different ways of using light have been developed. Some are used to save energy, for instance, solar cells, which use light to make electricity. The laser machine gives out a strong, narrow beam of light, which (unlike ordinary light) hardly

The strong, narrow beams of laser light are made up of a single color. These lasers at a nightclub create a dazzling display. Lasers also have many important uses in hospitals, stores, and factories.

spreads out. Laser beams are used in factories for cutting steel, in shops for reading bar codes on labels, at home with CD players, and in hospitals during delicate surgery. Light is also used to send messages down strands of coated glass called *optical fibers*. Optical fibers channel light in the same way that a pipe channels water. The message is sent as a coded series of flashing light: it can be a television program, computer information, or simply a phone call.

INVESTIGATION

Can you shine light along bends?

MATERIALS

A powerful flashlight, a large plastic bottle with a screw top, and water.

INSTRUCTIONS

Ask an adult to make a small hole in the side of the plastic bottle.

Place the bottle in a sink or a large bowl. Fill the bottle with water and screw on the top. The water will only flow out of the hole when you unscrew the cap.

Darken the room as much as possible and slowly unscrew the cap. Ask a friend to shine the flashlight on the bottle, opposite the hole, while you place a finger in the flow of water.

Notice the small circle of light on the end of your finger: this has traveled along the stream of water. Stop the flow of water and the small circle of light disappears.

FURTHER INVESTIGATION

Many scientific instruments have been given names made by combining two Ancient Greek words. Find the names of five instruments from this list of Greek words.

tele- (far off)
peri- (around)
micro- (small)
-scope (look at)
endo- (inside)
stetho- (chest)

These optical fibers are made of glass and carry light. Using optical fibers, it is possible to channel light around corners.

Glossary

Concave
Curving inward like a lens or mirror that is thinner in the middle than at the edges.

Convex
Curving outward like a lens or mirror that is thicker in the middle than at the edges.

Cornea
The transparent covering at the front of the eye.

Endoscope
An instrument for looking inside the body.

Focus
The point at which light rays meet after being reflected or refracted.

Illusion
Something that "tricks" the eye and the brain into believing what is not real.

Image
The picture of an object as seen in a mirror or through a lens.

Laser
A machine that produces a narrow beam of light of a particular color.

Lens
Any transparent object with a curved surface that bends light rays.

Microscope
An instrument that makes very small things look bigger.

Opaque
Used to describe an object that does not allow light to be passed through.

Optical fibers
Hair-fine glass rods that carry light.

Periscope
An instrument that uses mirrors so that we can see things that are above us, below us, or around a corner.

Pigments
Substances that color natural and artificially made objects.

Prism
A triangular piece of glass used to split light into the colors of the spectrum.

Pupil
The opening in the center of the eye where light enters.

Reflection
The name given to both the image of an object on a shiny surface and also to the bouncing back of light from a surface.

Refraction
The bending of light as it passes from one transparent surface to another.

Retina
The layer at the back of the eye that is sensitive to light.

Shadow
A dark region behind an object where it blocks light.

Solar
To do with the Sun.

Spectrum
The range of colors that make up white light.

Translucent
Translucent materials mix and scatter light waves as they pass through. This means you do not get a clear view through the material, only a vague image.

Transparent
Transparent materials let almost all the light waves through. Objects can be seen clearly through them.

Further information

BOOKS

DK Science Encyclopedia
(DK Children, 1999)

Eye-popping Optical Illusions
by Michael DiSpezio
(Sterling, 2002)

Light (Discover Science)
by Kim Taylor
(Chrysalis Education, 2003)

Light (Science Files)
by Steve Parker
(Heinemann Library, 2005)

Light, Color, and Art Activities (Arty Facts)
by Barbara Taylor
(Crabtree Publishing Company, 2002)

Light and Sound (The Young Oxford Library of Science)
by Jonathan Allday
(Oxford University Press, USA, 2003)

Simple Optical Illusion Experiments with Everyday Materials
by Michael DiSpezio
(Sterling, 2000)

CD-ROMS

Eyewitness Encyclopedia of Science
Global Software Publishing

I Love Science!
Global Software Publishing

ANSWERS

page 5 Different colors reflect different amounts of light so some are easier to see than others.

page 7 The shadow becomes smaller as the brick is moved back.

page 9 The image of the dot has fallen on your "blind spot" where the optic nerve joins the retina.

page 13 Each letter of CHOICE looks the same when it is written upside down. The letters of BALL do not.

page 14 Convex mirrors make you look shorter and concave mirrors make you look taller.

page 15 A dentist uses a concave mirror to look at our teeth.

The length of the smallest mirror would be about 3 ft. (90 cm)—half the height of the man who is 6 ft. (180 cm) tall.

A spoon. The bowl of a spoon curves inward (concave). The back of a spoon curves outward (convex).

page 17 Water can be deeper than it looks.

page 21 Blue and green light mix to make cyan (turquoise); green and red light mix to make yellow; red and blue light mix to make magenta (purplish-red). All three colors mix to make an off-white color.

The colors combine as follows: red and blue—purple; red and green—yellow; blue and green—cyan.

page 25 Pictures are upside down when they pass through a hole, because light travels in straight lines. Light from the top of an object goes to the bottom on the other side of the hole, and light from the bottom goes to the top.

A rectangular film fits into the circular image made by the lens.

page 27 The image of the window is smaller than the window itself, beacuse the light has passed through a convex lens (the magnifying glass).

The picture got smaller as you moved closer to the window.

The Sun's rays are focused through the magnifying glass and can start a fire.

page 29 The names of the instruments are as follows: telescope, periscope, microscope, endoscope, and stethoscope.

Index